20

I0466325

HABITABLE PLANETS

DISCOVERING NEW HOMES FOR HUMANITY

By Rayan Bale

Your Gateway to the Stars

TABLE OF CONTENTS

Introduction

Overview of Exoplanet Discovery

The discovery of exoplanets—planets that orbit stars outside our solar system—has revolutionized our understanding of the universe. It wasn't until the 1990s that the first exoplanet was confirmed, a breakthrough that opened the floodgates to an era of unprecedented astronomical discovery. Using advanced telescopes and sophisticated techniques, scientists have since identified thousands of exoplanets, revealing a diverse array of worlds that challenge our preconceived notions about planetary systems.

The Kepler Space Telescope, launched by NASA in 2009, marked a significant milestone in this journey. By continuously monitoring over 150,000 stars for subtle dips in brightness caused by transiting planets, Kepler alone has confirmed more than 2,600 exoplanets. Missions like TESS (Transiting Exoplanet Survey Satellite) and the upcoming James Webb Space Telescope promise to further expand our catalog of exoplanets, providing deeper insights into their characteristics and potential habitability.

Importance of Finding Habitable Planets

The search for habitable planets is not just about

finding new worlds; it is fundamentally about seeking potential homes for future human generations and answering profound questions about our place in the cosmos. The discovery of habitable exoplanets could provide critical answers to the age-old question of whether we are alone in the universe. It also has practical implications for humanity's long-term survival, as Earth faces increasing environmental challenges and the inevitability of cosmic events that could threaten life here.

Finding planets that could support life opens up the possibility of future colonization, ensuring the continuity of human civilization. Moreover, studying these distant worlds helps us understand the conditions that foster life, offering valuable lessons about the resilience and adaptability of life in the universe.

Brief Explanation of Criteria for Habitability

Determining whether an exoplanet is habitable involves assessing several key criteria. The primary factor is its location in the habitable zone of its star, often referred to as the "Goldilocks Zone." This is the region around a star where conditions are just right—not too hot and not too cold—for liquid water to exist on the planet's surface. Water is essential for life as we know it, making this a crucial requirement.

Another critical criterion is the planet's atmosphere. A stable and suitable atmosphere can regulate surface temperatures, protect the planet from harmful radiation, and support the chemical processes necessary for life. Scientists look for signs of water vapor, oxygen, carbon dioxide, and other gases that might indicate a life-sustaining environment.

The planet's size and composition also play significant roles. Earth-sized planets with rocky surfaces are more likely to support life than gas giants. Additionally, the activity level of the host star is considered; quieter stars provide more stable environments for potential life to develop and thrive.

Understanding these criteria helps scientists narrow down the vast number of exoplanets to those that offer the best chances for habitability. As our technology and methods improve, we are getting closer to finding worlds that might one day be new homes for humanity or reveal that life exists elsewhere in the universe.

Planet 1 :

Proxima Centauri b

Proxima Centauri b is the closest known exoplanet to Earth, situated just 4.24 light-years away in the Alpha Centauri star system. This intriguing planet orbits the red dwarf star Proxima Centauri, making it a focal point in the search for habitable worlds beyond our solar system.

Habitability Criteria:
One of the most compelling aspects of Proxima Centauri b is its location within the habitable zone of its parent star. This zone, often referred to as the "Goldilocks Zone," is the region around a star where conditions might be just right for liquid water to exist —a crucial ingredient for life as we know it.

Despite this promising position, the habitability of Proxima Centauri b faces significant challenges. The planet's star, Proxima Centauri, is a red dwarf that exhibits intense stellar flares. These flares can potentially strip away the planet's atmosphere, making it less hospitable for life. However, if Proxima Centauri b has a robust magnetic field or a thick atmosphere, it could protect itself from these harmful stellar emissions.

Research and Missions:
Proxima Centauri b was discovered by the European

Southern Observatory using the radial velocity method, which detects wobbles in a star's motion caused by the gravitational pull of an orbiting planet. Since its discovery, Proxima Centauri b has been the subject of extensive study, particularly regarding its atmospheric composition and potential to support life.

Researchers are keen to determine whether Proxima Centauri b has an atmosphere and, if so, what it is composed of. The presence of an atmosphere would significantly enhance the planet's potential habitability, providing a buffer against stellar flares and a medium for water and other life-sustaining compounds.

Expert Insights:

Leading astronomers have highlighted the importance of Proxima Centauri b in the ongoing search for habitable exoplanets. Dr. Guillem Anglada-Escudé, who led the team that discovered the planet, remarked, "Finding a potentially habitable planet so close to Earth is an extraordinary step forward in our quest to answer the age-old question: Are we alone in the universe?"

Future Prospects:

Looking ahead, Proxima Centauri b remains a prime

candidate for future interstellar missions. Concepts like Breakthrough Starshot propose sending tiny, lightweight probes to the Alpha Centauri system at a fraction of the speed of light, which could potentially reach Proxima Centauri b within a few decades. These missions aim to gather high-resolution images and direct data about the planet's atmosphere, surface conditions, and potential biosignatures.

Proxima Centauri b represents a tantalizing possibility in our search for new homes beyond Earth. Its proximity offers a unique opportunity to study and possibly even visit a potentially habitable exoplanet within our lifetimes, marking a significant milestone in humanity's journey to explore the cosmos.

Planet 2 : TRAPPIST-1e

TRAPPIST-1e is a fascinating exoplanet located 39 light-years away in the TRAPPIST-1 system. This system is notable for its seven Earth-sized planets, three of which are within the habitable zone. TRAPPIST-1e, being one of these three, has garnered significant interest due to its potential to support life.

Habitability Criteria:

TRAPPIST-1e resides comfortably within the habitable zone of its parent star, a cool red dwarf named TRAPPIST-1. This positioning suggests that the planet could have the right conditions for liquid water to exist on its surface. Additionally, the planet is believed to have a stable atmosphere, which is crucial for maintaining surface conditions that could support life.

The presence of liquid water and a stable atmosphere are key indicators of habitability. If these conditions are confirmed, TRAPPIST-1e would be a prime candidate in the search for extraterrestrial life. The planet's relatively low stellar activity compared to other red dwarfs further enhances its potential habitability.

Research and Missions:

TRAPPIST-1e was discovered by the TRAPPIST

(Transiting Planets and Planetesimals Small Telescope) project and has been extensively studied by NASA's Spitzer Space Telescope. These observations have provided critical data on the planet's size, orbit, and potential atmospheric properties.

The extensive research on TRAPPIST-1e's atmospheric conditions aims to determine whether the planet can support liquid water. Initial findings suggest that the planet could indeed possess the necessary elements for habitability, but more detailed studies are required to confirm these hypotheses.

Expert Insights:
Researchers involved in the discovery and study of TRAPPIST-1e have expressed great excitement about its potential. Dr. Michaël Gillon, the lead scientist of the TRAPPIST project, noted, "The TRAPPIST-1 system is like a treasure chest of habitable worlds. TRAPPIST-1e, in particular, stands out as a promising candidate for further exploration."

Future Prospects:
The future of TRAPPIST-1e's exploration looks promising, especially with the advent of the James Webb Space Telescope (JWST). The JWST is

expected to provide unprecedented detail in the study of exoplanet atmospheres, including TRAPPIST-1e. Researchers anticipate using the JWST to conduct detailed atmospheric analysis, which could reveal the presence of water vapor, oxygen, methane, and other potential biosignatures.

TRAPPIST-1e continues to be a focal point in the search for habitable exoplanets. Its potential to support life makes it an exciting target for future missions and studies, offering a glimpse into the possibility of finding another Earth-like world within our galaxy.

Planet 3 :
Kepler-186f

Kepler-186f is a remarkable exoplanet located 492 light-years away in the constellation Cygnus. It holds the distinction of being the first Earth-sized planet discovered within the habitable zone of another star, making it a groundbreaking find in the quest to locate potentially habitable worlds.

Habitability Criteria:
Kepler-186f receives a similar amount of stellar energy from its host star as Earth does from the Sun, placing it within the habitable zone where conditions could allow for the presence of liquid water. The planet's size and energy receipt suggest that it could have surface conditions conducive to supporting life.

Although being in the habitable zone is a significant factor, the potential for liquid water on Kepler-186f also depends on its atmospheric composition and pressure. The presence of a stable and sufficient atmosphere would be crucial to maintaining surface temperatures that allow water to remain in liquid form.

Research and Missions:
Kepler-186f was discovered by the Kepler Space Telescope, which used the transit method to detect the planet as it passed in front of its host star, causing a slight dimming in the star's light. This

method provided data on the planet's size and orbit, indicating its potential habitability.

Subsequent studies have focused on understanding the planet's composition and atmosphere. While direct observations of its atmosphere are challenging due to its distance, models and simulations suggest that Kepler-186f could have an atmosphere capable of supporting liquid water, assuming it has a suitable mix of gases and surface pressure.

Expert Insights:
Kepler mission scientists have emphasized the importance of Kepler-186f in the search for habitable exoplanets. Dr. Elisa Quintana, a research scientist at NASA's Ames Research Center, stated, "The discovery of Kepler-186f is a significant step toward finding worlds like our own. It opens up a new frontier in the search for life beyond Earth."

Future Prospects:
Further observation of Kepler-186f is needed to confirm its habitability. Future missions and telescopes, such as the James Webb Space Telescope, may provide the capability to analyze the planet's atmosphere in greater detail. Understanding the atmospheric composition and detecting potential biosignatures would be crucial in assessing the

planet's ability to support life.

 Kepler-186f remains a beacon of hope in the search for Earth-like planets. Its discovery has paved the way for more detailed studies of similar exoplanets, bringing us closer to answering the profound question of whether we are alone in the universe. The continued exploration of Kepler-186f will undoubtedly yield more insights into the potential for life on planets beyond our solar system.

Planet 4 :

Kepler-442b

Kepler-442b is an intriguing exoplanet located 1,206 light-years away from Earth. It orbits a K-type star, which is smaller and cooler than our Sun. This distant world has garnered attention due to its promising position within its star's habitable zone, where conditions might support liquid water and, potentially, life.

Habitability Criteria:

Kepler-442b resides within the habitable zone of its host star, meaning it is at a distance where temperatures could allow for the presence of liquid water on its surface. The planet receives about 70% of the light that Earth receives from the Sun, which is sufficient to maintain moderate temperatures conducive to habitability.

The possibility of liquid water is a key factor in assessing the habitability of Kepler-442b. If the planet has a suitable atmosphere, it could maintain the right surface pressure and temperature for water to exist in liquid form, providing an essential ingredient for life as we know it.

Research and Missions:

Kepler-442b was discovered by the Kepler Space Telescope using the transit method, which involves detecting the slight dimming of a star as a planet passes in front of it. This method provided crucial data on the planet's size and orbit, helping researchers determine its potential habitability.

Subsequent studies have focused on modeling the planet's atmosphere and surface conditions. These studies aim to predict whether Kepler-442b has the necessary atmospheric composition to support liquid water and sustain a stable climate over long periods.

Expert Insights:

Researchers have expressed optimism about Kepler-442b's potential for habitability. Dr. Jeffrey Coughlin, a Kepler research scientist, stated, "Kepler-442b is one of the best candidates we have for a potentially habitable planet. Its size and position within the habitable zone make it a prime target for future observations."

Future Prospects:

Continued observation of Kepler-442b is essential to confirm its habitability. Future missions and advanced telescopes, such as the James Webb Space Telescope, are expected to provide more detailed data on the planet's atmosphere and surface conditions. These observations will help determine the presence of water vapor, greenhouse gases, and other markers that could indicate the planet's potential to support life.

Kepler-442b remains a significant candidate in the search for habitable exoplanets. Its discovery has opened new avenues for exploring worlds beyond our solar system, and future observations will further our understanding of its capacity to host life. The study of Kepler-442b continues to inspire and drive the quest to find new homes for humanity in the cosmos.

Planet 5 :

LHS 1140 b

LHS 1140 b is an exciting exoplanet located 40 light-years away in the constellation Cetus. It orbits a red dwarf star, making it a subject of interest in the search for habitable worlds due to its proximity and potential for supporting life.

Habitability Criteria:
LHS 1140 b is situated within the habitable zone of its parent star, a region where conditions could allow for the presence of liquid water on the planet's surface. The planet is believed to have a dense atmosphere, which is crucial for maintaining stable surface temperatures and protecting against stellar radiation.

The combination of being in the habitable zone and having a dense atmosphere increases the likelihood that LHS 1140 b could harbor liquid water, an essential component for life as we know it. These factors make it a prime candidate for further study in the quest to find habitable exoplanets.

Research and Missions:
LHS 1140 b was discovered by the MEarth Project, which focuses on finding Earth-like planets around nearby red dwarf stars. The discovery was made using the transit method, where the planet's passage in front of its star causes a detectable dip in the star's

brightness.

Subsequent studies have concentrated on determining the planet's size, density, and atmospheric composition. These studies suggest that LHS 1140 b has a rocky composition with a high density, indicating it could have a substantial atmosphere capable of supporting liquid water.

Expert Insights:
Astronomers have expressed optimism about the habitability of LHS 1140 b. Dr. Jason Dittmann, who led the discovery team, remarked, "LHS 1140 b is one of the most exciting exoplanets we've found in recent years. Its size, density, and position within the habitable zone make it an excellent candidate for further study."

Future Prospects:
LHS 1140 b is a key target for detailed atmospheric study by the James Webb Space Telescope (JWST). The JWST's advanced capabilities will allow scientists to analyze the planet's atmosphere in greater detail, searching for signs of water vapor, greenhouse gases, and other indicators of habitability.

The potential for LHS 1140 b to support life makes it

a focal point for future missions and observations. Understanding its atmosphere and surface conditions will provide critical insights into the planet's ability to sustain life, contributing to the broader search for new homes for humanity beyond our solar system.

LHS 1140 b represents a promising opportunity in the exploration of potentially habitable exoplanets. Its favorable location and characteristics inspire hope and excitement in the scientific community, driving the continued pursuit of discovering life beyond Earth.

Planet 6 :
Gliese 667 Cc

Gliese 667 Cc is an intriguing exoplanet located 23.62 light-years away in the constellation Scorpius. This planet orbits within the habitable zone of its host star, making it a prominent candidate in the search for potentially habitable worlds.

Habitability Criteria:

Gliese 667 Cc receives a similar amount of stellar energy as Earth, which places it in the habitable zone where conditions might be right for liquid water to exist on its surface. The potential for liquid water is a crucial factor in assessing the planet's habitability, as it is a fundamental ingredient for life as we know it.

The planet's position in the habitable zone, coupled with its potential to maintain liquid water, makes it a significant target for researchers studying exoplanetary habitability. Its stable energy reception from its star suggests that Gliese 667 Cc could have a climate capable of supporting life.

Research and Missions:

Gliese 667 Cc was discovered by the European Southern Observatory using the radial velocity method, which detects variations in a star's motion caused by the gravitational pull of an orbiting planet. This method provided essential data on the planet's orbit and size, helping scientists evaluate its potential habitability.

Subsequent studies have focused on understanding the planet's orbit, size, and the conditions that might prevail on its surface. Researchers are particularly interested in the planet's atmosphere and the likelihood of it supporting liquid water.

Expert Insights:
Scientists have emphasized the importance of Gliese 667 Cc in the context of exoplanet research. Dr. Guillem Anglada-Escudé, who has studied the Gliese 667 system extensively, commented, "Gliese 667 Cc is one of the best candidates for a potentially habitable planet. Its location in the habitable zone and the energy it receives from its star make it an exciting subject for further investigation."

Future Prospects:
There is considerable interest in continuing to study and observe Gliese 667 Cc. Future missions and advanced telescopes will aim to gather more detailed data on the planet's atmosphere and surface conditions. Instruments such as the James Webb Space Telescope (JWST) could provide valuable insights into the atmospheric composition and potential biosignatures of Gliese 667 Cc.

The study of Gliese 667 Cc represents a promising avenue for understanding the conditions necessary for habitability on exoplanets. As technology and observational capabilities advance, scientists hope to learn more about this intriguing world and its potential to support life.

Gliese 667 Cc stands as a significant milestone in the quest to find new homes for humanity. Its promising characteristics and proximity to Earth make it an essential target for future research, inspiring continued efforts to explore the possibilities of life beyond our solar system.

Planet 7 :
K2-18b

K2-18b is a captivating exoplanet located 124 light-years away in the constellation Leo. This planet orbits a red dwarf star and has drawn significant attention due to its promising habitability characteristics, particularly the presence of water vapor in its atmosphere.

Habitability Criteria:
K2-18b lies within the habitable zone of its host star, where conditions might allow for the existence of liquid water on its surface. The discovery of water vapor in its atmosphere is a crucial indicator of potential habitability, as water is essential for life as we know it.

The combination of being in the habitable zone and having detected water vapor suggests that K2-18b could have a climate capable of supporting life. This finding significantly enhances its status as a candidate for further study in the search for habitable exoplanets.

Research and Missions:
K2-18b was discovered by the K2 mission of the Kepler Space Telescope using the transit method. This method involves detecting the slight dimming of a star as the planet passes in front of it, providing data on the planet's size and orbit.

Subsequent studies have focused on analyzing the atmospheric composition of K2-18b. Observations using the Hubble Space Telescope revealed the presence of water vapor in the planet's atmosphere, marking a significant milestone in exoplanet research.

Expert Insights:

Researchers have highlighted the importance of the water vapor discovery on K2-18b. Dr. Angelos Tsiaras, who led the team that made the discovery, remarked, "The detection of water vapor in the atmosphere of K2-18b is a major breakthrough. It brings us one step closer to understanding the habitability of planets beyond our solar system."

Future Prospects:

K2-18b is a prime target for detailed atmospheric analysis by future missions and advanced telescopes. The James Webb Space Telescope (JWST) will play a crucial role in providing more detailed data on the planet's atmosphere, including its composition and the presence of potential biosignatures.

The study of K2-18b will continue to focus on understanding its atmospheric properties and surface conditions. Future observations aim to confirm the presence of other vital elements and compounds that could support life, making K2-18b an essential subject in the ongoing search for habitable exoplanets.

 K2-18b represents a significant step forward in our quest to find new homes for humanity. Its favorable location, promising atmospheric conditions, and the discovery of water vapor inspire hope and excitement within the scientific community, driving further exploration and study of this intriguing exoplanet.

Planet 8 :
HD 40307g

HD 40307g is an intriguing exoplanet located 42 light-years away in the constellation Pictor. It orbits a K-type star, which is cooler and smaller than our Sun, and has garnered interest due to its potential habitability.

Habitability Criteria:
HD 40307g is situated within the habitable zone of its host star, where conditions could allow for the presence of liquid water on its surface. The planet's position in the habitable zone suggests it might have a stable climate capable of supporting life.

The potential for a stable climate and liquid water makes HD 40307g a significant candidate for habitability. If the planet has a suitable atmosphere, it could maintain the right surface conditions for life to thrive.

Research and Missions:
HD 40307g was discovered by the High Accuracy Radial Velocity Planet Searcher (HARPS) using the radial velocity method. This method detects variations in a star's motion caused by the gravitational pull of an orbiting planet, providing crucial data on the planet's orbit and mass.

Subsequent studies have focused on understanding

HD 40307g's orbit and the conditions that might prevail on its surface. Researchers are particularly interested in the planet's potential to maintain liquid water and a stable climate over long periods.

Expert Insights:

Astronomers have shared their insights on the habitability potential of HD 40307g. Dr. Mikko Tuomi, who has studied the planet, noted, "HD 40307g is one of the more promising candidates for habitability. Its location in the habitable zone and the characteristics of its host star make it an exciting subject for further study."

Future Prospects:

Further observation of HD 40307g is essential to confirm its habitability. Future missions and advanced telescopes, such as the James Webb Space Telescope (JWST), are expected to provide more detailed data on the planet's atmosphere and surface conditions. These observations will help determine the presence of water vapor, greenhouse gases, and other markers that could indicate the planet's potential to support life.

HD 40307g remains a significant candidate in the search for habitable exoplanets. Its promising characteristics and relatively close proximity to Earth make it an important target for future research. Continued study of HD 40307g will enhance our understanding of the conditions necessary for habitability and contribute to the broader quest for new homes for humanity beyond our solar system.

The exploration of HD 40307g represents a critical step in uncovering the mysteries of potentially habitable worlds. As technology advances and our observational capabilities improve, we look forward to discovering more about this intriguing exoplanet and its potential to support life.

Planet 9 :

Wolf 1061c

Wolf 1061c is a captivating exoplanet situated 13.8 light-years away in the constellation Ophiuchus. It orbits a red dwarf star and has become a focal point for researchers due to its potential habitability and relative proximity to Earth.

Habitability Criteria:

Wolf 1061c is located within the habitable zone of its host star, where conditions might allow for the presence of liquid water on its surface. This positioning suggests that the planet could have surface temperatures suitable for sustaining life.

The potential for liquid water is a key factor in assessing Wolf 1061c's habitability. If the planet possesses a stable atmosphere, it could maintain the right surface pressure and temperature to support water in liquid form, providing an essential ingredient for life as we know it.

Research and Missions:

Wolf 1061c was discovered by the High Accuracy Radial Velocity Planet Searcher (HARPS) using the radial velocity method. This method detects variations in a star's motion caused by the gravitational pull of an orbiting planet, allowing scientists to determine the planet's size, orbit, and mass.

Subsequent studies have focused on understanding the planet's characteristics and its potential for habitability. Researchers have been particularly interested in its orbit and the conditions that might prevail on its surface, seeking to determine if Wolf 1061c could support liquid water and, by extension, life.

Expert Insights:

Scientists have expressed optimism about Wolf 1061c's potential habitability. Dr. Duncan Wright, who has studied the planet, stated, "Wolf 1061c is one of the closest potentially habitable planets we know of. Its position in the habitable zone and the characteristics of its star make it an exciting candidate for further study."

Future Prospects:

There is considerable interest in continuing to study and observe Wolf 1061c. Future missions and advanced telescopes, such as the James Webb Space Telescope (JWST), will provide more detailed data on the planet's atmosphere and surface conditions. These observations will help determine the presence of water vapor, greenhouse gases, and other indicators of habitability.

The study of Wolf 1061c will focus on confirming its

potential to support life. Understanding its atmospheric composition and surface conditions will provide critical insights into the planet's capacity for habitability, contributing to the broader search for new homes for humanity beyond our solar system.

Wolf 1061c remains a significant candidate in the quest to find habitable exoplanets. Its promising characteristics and close proximity to Earth make it an essential target for future research, inspiring continued efforts to explore the possibilities of life beyond our solar system. The exploration of Wolf 1061c represents a critical step in uncovering the mysteries of potentially habitable worlds and the potential for life beyond our planet.

Planet 10 :
Ross 128 b

Ross 128 b is an intriguing exoplanet located just 11 light-years away in the constellation Virgo. It orbits a relatively quiet red dwarf star, making it an appealing candidate in the search for habitable worlds due to the stable stellar environment and its proximity to Earth.

Habitability Criteria:

Ross 128 b is situated within the habitable zone of its host star, meaning it is at a distance where temperatures could allow for the presence of liquid water on the planet's surface. This positioning, along with the quiet nature of its parent star, suggests that Ross 128 b could have a stable climate conducive to life.

The potential for a stable climate and liquid water enhances Ross 128 b's status as a promising candidate for habitability. If the planet has a suitable atmosphere, it could maintain the right surface conditions to support life.

Research and Missions:

Ross 128 b was discovered by the High Accuracy Radial Velocity Planet Searcher (HARPS) using the radial velocity method. This technique detects variations in a star's motion caused by the gravitational pull of an orbiting planet, providing essential data on the planet's orbit and mass.

Subsequent studies have focused on understanding Ross 128 b's orbit, size, and atmospheric composition. Researchers are particularly interested in determining whether the planet has an atmosphere that could support liquid water and sustain a stable climate.

Expert Insights:
Researchers have shared positive perspectives on the habitability potential of Ross 128 b. Dr. Xavier Bonfils, who was involved in the discovery, remarked, "Ross 128 b is one of the closest and most promising habitable exoplanets we've discovered. Its quiet host star and location in the habitable zone make it an exciting target for further study."

Future Prospects:
Ross 128 b is a prime target for future observation and study. Advanced telescopes like the James Webb Space Telescope (JWST) will be instrumental in providing more detailed data on the planet's atmosphere and surface conditions. These observations will help determine the presence of water vapor, greenhouse gases, and other markers that could indicate the planet's potential to support life.

The study of Ross 128 b will focus on confirming its habitability potential. Understanding its atmospheric properties and surface conditions will provide critical insights into the planet's capacity for sustaining life, contributing to the broader search for new homes for humanity beyond our solar system.

Ross 128 b stands out as a significant candidate in the quest to find habitable exoplanets. Its promising characteristics, combined with its relatively close proximity to Earth, make it an essential target for future research. The exploration of Ross 128 b represents a crucial step in uncovering the mysteries of potentially habitable worlds and the potential for life beyond our planet.

Planet 11 :
Tau Ceti e

Tau Ceti e is an intriguing exoplanet located approximately 11.9 light-years away from Earth in the constellation Cetus. It orbits the star Tau Ceti, a G-type star similar to our Sun but slightly smaller and less luminous. Tau Ceti e is one of four known planets in the Tau Ceti system and has drawn significant interest due to its potential habitability and proximity to Earth.

Habitability Criteria:

Tau Ceti e is situated within the inner edge of its star's habitable zone, the region where conditions might allow for liquid water to exist on the planet's surface. Although it receives more stellar radiation than Earth, its position suggests that with the right atmospheric conditions, it could support temperatures conducive to liquid water. The planet's size is also a critical factor; it is slightly larger than Earth, which could imply a rocky composition, although this needs to be confirmed.

Research and Missions:

Tau Ceti e was discovered using the radial velocity method, which detects variations in a star's motion caused by the gravitational pull of orbiting planets. This technique has been employed by several observatories, including the European Southern Observatory (ESO) and other international teams.

The discovery of Tau Ceti e and its sibling planets was part of efforts to identify potentially habitable worlds around nearby stars.

Subsequent studies have focused on modeling the planet's climate and atmospheric conditions to better understand its potential for habitability. These models consider various scenarios, including different atmospheric compositions and greenhouse gas levels, to estimate surface temperatures and the likelihood of liquid water.

Expert Insights:
Scientists studying Tau Ceti e have emphasized the importance of understanding the planet's atmosphere to assess its habitability. Dr. Mikko Tuomi, who has been involved in the study of the Tau Ceti system, stated, "Tau Ceti e is an excellent candidate for studying the conditions necessary for habitability. Its proximity to Earth makes it an ideal target for future missions and observations."

Future Prospects:
Future observations of Tau Ceti e will aim to determine its atmospheric composition and surface conditions more precisely. The upcoming James Webb Space Telescope (JWST) and other advanced observatories will provide critical data that could

confirm the presence of water vapor, greenhouse gases, and other markers of habitability.

Additionally, missions like the European Space Agency's PLATO and ground-based projects such as the Extremely Large Telescope (ELT) will contribute to our understanding of Tau Ceti e and its potential to support life. These instruments will offer more detailed observations and help refine models of the planet's climate and atmospheric dynamics.

Tau Ceti e represents a promising candidate in the search for habitable exoplanets. Its proximity to Earth and the similarities of its host star to our Sun make it a compelling target for ongoing and future research. The study of Tau Ceti e will not only enhance our understanding of exoplanetary systems but also contribute to the broader quest to find new homes for humanity in the cosmos.

Planet 12 :
Teegarden b

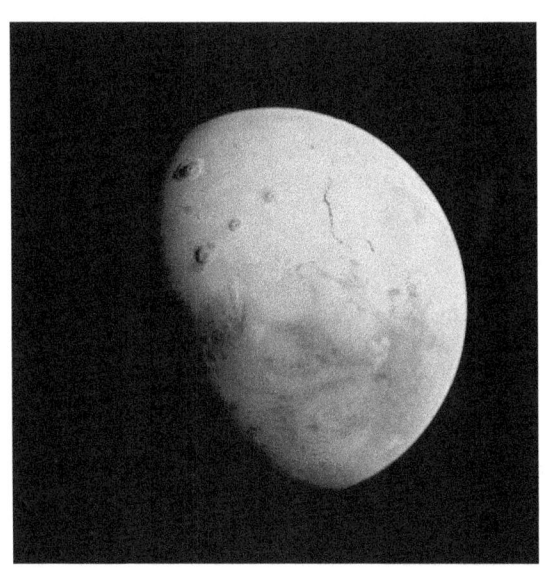

Teegarden b is an intriguing exoplanet located approximately 12 light-years away from Earth in the constellation Aries. It orbits the red dwarf star Teegarden's Star, which is much cooler and smaller than our Sun. Teegarden b is one of two planets discovered in this system and has garnered significant interest due to its proximity and potential habitability.

Habitability Criteria:
Teegarden b is situated within the habitable zone of its host star, where conditions might allow for liquid water to exist on its surface. The planet is believed to have a mass similar to Earth's, which suggests a rocky composition. Given its location in the habitable zone and its Earth-like size, Teegarden b has the potential to maintain surface temperatures conducive to life, provided it has a suitable atmosphere.

Research and Missions:
Teegarden b was discovered by the CARMENES project, which uses the radial velocity method to detect exoplanets. This method involves observing the star's motion to infer the presence of orbiting planets. The discovery of Teegarden b was part of a broader effort to find Earth-like planets around nearby red dwarf stars.

Studies have since focused on understanding the planet's characteristics and potential habitability. These include modeling the planet's climate and atmospheric conditions to predict surface temperatures and the possibility of liquid water. Given its proximity to Earth, Teegarden b is a prime candidate for future observational campaigns.

Expert Insights:

Researchers have highlighted the significance of Teegarden b in the context of exoplanetary studies. Dr. Mathias Zechmeister, a key figure in the discovery, stated, "Teegarden b is one of the closest known exoplanets that could have conditions suitable for life. Its proximity allows us to study it in greater detail and opens up exciting possibilities for future exploration."

Future Prospects:

The future of Teegarden b's exploration looks promising, particularly with the advent of new telescopes and missions. The James Webb Space Telescope (JWST) is expected to provide detailed observations of the planet's atmosphere, searching for signs of water vapor, greenhouse gases, and other indicators of habitability.

Additionally, the European Space Agency's PLATO

mission and the ground-based Extremely Large Telescope (ELT) will offer further insights into Teegarden b's atmospheric composition and surface conditions. These advanced instruments will help confirm whether the planet has the necessary conditions to support life.

Teegarden b remains a significant candidate in the search for habitable exoplanets. Its proximity to Earth and potential for habitability make it an essential target for ongoing and future research. As we continue to explore this fascinating planet, we move closer to understanding the conditions necessary for life beyond our solar system and the possibility of finding new homes for humanity among the stars.

Planet 13 :

Kapteyn b

Kapteyn b is a fascinating exoplanet located approximately 12.8 light-years away from Earth in the constellation Pictor. It orbits the red subdwarf star Kapteyn's Star, which is much older and less luminous than our Sun. Kapteyn b is notable for being one of the oldest known potentially habitable exoplanets, estimated to be around 11 billion years old.

Habitability Criteria:

Kapteyn b resides within the habitable zone of its host star, where conditions might allow for the presence of liquid water on its surface. The planet has a minimum mass of about 4.8 times that of Earth, suggesting it could be a super-Earth with a rocky composition. Given its location in the habitable zone and its size, Kapteyn b could maintain surface temperatures suitable for life if it has a stable atmosphere.

Research and Missions:

Kapteyn b was discovered by the High Accuracy Radial Velocity Planet Searcher (HARPS) using the radial velocity method. This method detects variations in a star's motion caused by the gravitational pull of orbiting planets. The discovery of Kapteyn b, along with its sibling planet Kapteyn c, was part of an effort to identify potentially habitable planets around nearby stars.

Subsequent studies have focused on modeling the planet's climate and atmospheric conditions to better understand its potential habitability. Researchers have considered various scenarios, including different atmospheric compositions and greenhouse gas levels, to estimate surface temperatures and the likelihood of liquid water.

Expert Insights:
Scientists studying Kapteyn b have emphasized the planet's unique characteristics due to its age and potential habitability. Dr. Guillem Anglada-Escudé, who led the discovery team, remarked, "Kapteyn b is a remarkable find, not just because of its potential habitability but also due to its ancient age. It provides a unique opportunity to study the conditions for habitability over an extended period."

Future Prospects:
Future observations of Kapteyn b will aim to determine its atmospheric composition and surface conditions more precisely. Advanced telescopes like the James Webb Space Telescope (JWST) and the European Space Agency's PLATO mission will be crucial in providing detailed data that could confirm the presence of water vapor, greenhouse gases, and

other markers of habitability.

The Extremely Large Telescope (ELT) and other ground-based observatories will also contribute to our understanding of Kapteyn b. These instruments will offer more detailed observations and help refine models of the planet's climate and atmospheric dynamics.

Kapteyn b represents a promising candidate in the search for habitable exoplanets. Its unique characteristics, including its ancient age and location in the habitable zone, make it an important target for future research. The study of Kapteyn b will enhance our understanding of the conditions necessary for habitability and contribute to the broader quest to find new homes for humanity beyond our solar system.

As we continue to explore this intriguing planet, we gain insights into the longevity of habitable conditions and the potential for life to persist over billions of years. Kapteyn b stands as a testament to the enduring nature of the search for life beyond Earth and the possibilities that lie within our galaxy.

Planet 14 :
GJ 667 C f

GJ 667 C f is an exoplanet located approximately 23.62 light-years away from Earth in the constellation Scorpius. It orbits one of the three stars in the GJ 667 system, specifically the M-class red dwarf star GJ 667 C. This planet is part of a complex system with multiple planets, some of which are located within the habitable zone.

Habitability Criteria:
GJ 667 C f is situated within the habitable zone of its host star, where conditions might allow for liquid water to exist on its surface. The planet's mass and size suggest it could be a super-Earth, which means it may have a rocky composition and the potential to support a stable atmosphere. The relatively low stellar activity of GJ 667 C enhances the prospects for a stable climate, making it a promising candidate for habitability.

Research and Missions:
GJ 667 C f was discovered using the radial velocity method by a team of astronomers led by the European Southern Observatory (ESO). This technique involves detecting variations in a star's motion caused by the gravitational pull of orbiting planets. The discovery of GJ 667 C f, along with other planets in the GJ 667 C system, has been a significant step in the study of potentially habitable

exoplanets.

Subsequent studies have focused on understanding the planet's characteristics and potential habitability. Researchers have modeled the planet's climate and atmospheric conditions to estimate surface temperatures and the possibility of liquid water. The GJ 667 C system's unique configuration allows for detailed comparative studies of multiple planets within the same system.

Expert Insights:

Astronomers have highlighted the importance of GJ 667 C f in exoplanetary research. Dr. Guillem Anglada-Escudé, a key figure in the study of the GJ 667 C system, noted, "The discovery of GJ 667 C f within the habitable zone is exciting because it adds to the growing list of potentially habitable planets around nearby stars. Its location in a multi-planet system allows us to study the dynamics and habitability of planets in such environments."

Future Prospects:

Future observations of GJ 667 C f will focus on determining its atmospheric composition and surface conditions. The James Webb Space Telescope (JWST) and the European Space Agency's PLATO mission are expected to provide critical data on the planet's atmosphere, including the presence of water vapor, greenhouse gases, and other markers of habitability.

Ground-based observatories such as the Extremely Large Telescope (ELT) will also play a crucial role in studying GJ 667 C f. These advanced instruments will offer more detailed observations and help refine models of the planet's climate and atmospheric dynamics.

GJ 667 C f remains a significant candidate in the search for habitable exoplanets. Its promising characteristics and location within a multi-planet system make it an essential target for future research. The study of GJ 667 C f will enhance our understanding of the conditions necessary for habitability and contribute to the broader quest to find new homes for humanity beyond our solar system.

As we continue to explore this intriguing planet, we gain valuable insights into the complexities of planetary systems and the potential for life to exist under diverse conditions. GJ 667 C f stands as a testament to the richness and variety of the cosmos, inspiring continued efforts to discover and study potentially habitable worlds.

Planet 15 :
Luyten b

Luyten b, also known as GJ 273b, is an exoplanet located approximately 12.2 light-years away from Earth in the constellation Canis Minor. It orbits the red dwarf star Luyten's Star (GJ 273), which is smaller and cooler than our Sun. Luyten b has attracted significant interest due to its proximity to Earth and its potential habitability.

Habitability Criteria:

Luyten b resides within the habitable zone of its host star, where conditions might allow for liquid water to exist on its surface. The planet has an estimated mass of about 2.9 times that of Earth, suggesting it could be a super-Earth with a rocky composition. Given its location in the habitable zone and its Earth-like characteristics, Luyten b holds promise for maintaining surface temperatures conducive to life, provided it has a suitable atmosphere.

Research and Missions:

Luyten b was discovered by the High Accuracy Radial Velocity Planet Searcher (HARPS) using the radial velocity method. This method detects variations in a star's motion caused by the gravitational pull of orbiting planets, providing critical data on the planet's orbit and mass. The discovery of Luyten b was part of efforts to identify potentially habitable planets around nearby stars.

Subsequent studies have focused on modeling the planet's climate and atmospheric conditions to better understand its potential habitability. Researchers have considered various scenarios, including different atmospheric compositions and greenhouse gas levels, to estimate surface temperatures and the likelihood of liquid water.

Expert Insights:
Scientists studying Luyten b have highlighted the planet's potential for habitability and its importance in exoplanet research. Dr. Xavier Bonfils, who has been involved in the study, remarked, "Luyten b is a fascinating candidate for habitability due to its proximity and favorable conditions. Its discovery opens up exciting possibilities for studying Earth-like planets around nearby stars."

Future Prospects:
Future observations of Luyten b will aim to determine its atmospheric composition and surface conditions more precisely. Advanced telescopes such as the James Webb Space Telescope (JWST) and the European Space Agency's PLATO mission will provide crucial data that could confirm the presence of water vapor, greenhouse gases, and other markers of habitability.

Ground-based observatories like the Extremely Large Telescope (ELT) will also contribute to our understanding of Luyten b. These instruments will offer more detailed observations and help refine models of the planet's climate and atmospheric dynamics.

Luyten b remains a significant candidate in the search for habitable exoplanets. Its promising characteristics and relatively close proximity to Earth make it an essential target for future research. The study of Luyten b will enhance our understanding of the conditions necessary for habitability and contribute to the broader quest to find new homes for humanity beyond our solar system.

As we continue to explore this intriguing planet, we gain insights into the potential for life to exist under diverse conditions and the complexities of planetary systems. Luyten b stands as a testament to the richness and variety of the cosmos, inspiring continued efforts to discover and study potentially habitable worlds.

Planet 16 :
Wolf 1061d

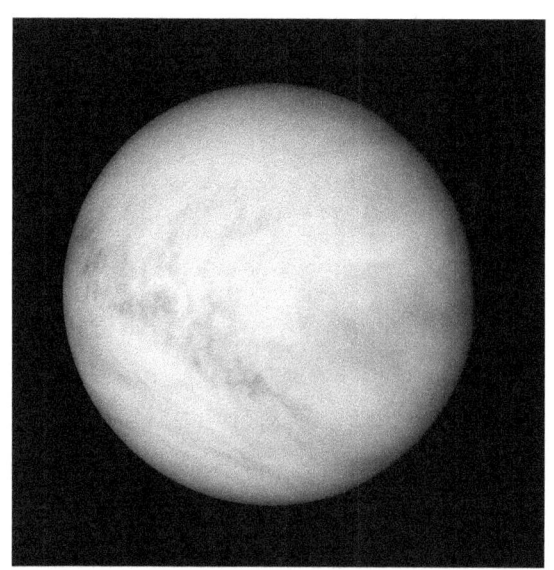

Wolf 1061d is an intriguing exoplanet located approximately 13.8 light-years away from Earth in the constellation Ophiuchus. It orbits the red dwarf star Wolf 1061, which is smaller and cooler than our Sun. Wolf 1061d is part of a system with at least three known planets, and it has drawn considerable attention due to its potential for habitability.

Habitability Criteria:

Wolf 1061d is situated at the outer edge of the habitable zone of its host star, meaning conditions might allow for liquid water to exist on its surface under certain circumstances. The planet has a mass approximately 7.5 times that of Earth, classifying it as a super-Earth. Its larger mass suggests it could have a significant atmosphere, which is essential for maintaining surface temperatures that could support liquid water.

Research and Missions:

Wolf 1061d was discovered by the High Accuracy Radial Velocity Planet Searcher (HARPS) using the radial velocity method. This technique involves detecting variations in a star's motion caused by the gravitational pull of orbiting planets, allowing scientists to infer the presence of the planet and determine its mass and orbit.

Subsequent studies have focused on modeling the planet's climate and atmospheric conditions to better understand its potential habitability. Researchers have explored various scenarios, including different atmospheric compositions and greenhouse gas levels, to estimate surface temperatures and the likelihood of liquid water.

Expert Insights:

Scientists have expressed cautious optimism about the habitability of Wolf 1061d. Dr. Stephen Kane, an astronomer who has studied the Wolf 1061 system, commented, "Wolf 1061d is a compelling candidate for further study due to its location at the edge of the habitable zone. Understanding the planet's atmosphere and surface conditions will be key to assessing its potential for habitability."

Future Prospects:

Future observations of Wolf 1061d will aim to determine its atmospheric composition and surface conditions more precisely. Advanced telescopes such as the James Webb Space Telescope (JWST) and the European Space Agency's PLATO mission will be instrumental in providing detailed data on the planet's atmosphere, including the presence of water vapor, greenhouse gases, and other markers of habitability.

Ground-based observatories like the Extremely Large Telescope (ELT) will also play a crucial role in studying Wolf 1061d. These advanced instruments will offer more detailed observations and help refine models of the planet's climate and atmospheric dynamics.

Wolf 1061d remains a significant candidate in the search for habitable exoplanets. Its promising characteristics and relatively close proximity to Earth make it an essential target for future research. The study of Wolf 1061d will enhance our understanding of the conditions necessary for habitability and contribute to the broader quest to find new homes for humanity beyond our solar system.

As we continue to explore this intriguing planet, we gain valuable insights into the complexities of planetary systems and the potential for life to exist under diverse conditions. Wolf 1061d stands as a testament to the richness and variety of the cosmos, inspiring continued efforts to discover and study potentially habitable worlds.

Planet 17 :
Kepler-62f

Kepler-62f is a fascinating exoplanet located approximately 1,200 light-years away from Earth in the constellation Lyra. It orbits the star Kepler-62, which is a K-type main-sequence star that is smaller and cooler than our Sun. Kepler-62f is part of a planetary system that includes five known planets, and it has garnered significant attention due to its potential habitability.

Habitability Criteria:

Kepler-62f resides within the habitable zone of its host star, where conditions might allow for liquid water to exist on its surface. The planet is about 1.4 times the size of Earth, which suggests that it could have a rocky composition. Its size also indicates that it may have a significant atmosphere capable of regulating surface temperatures. Given its position in the habitable zone and its Earth-like size, Kepler-62f is considered a promising candidate for maintaining conditions suitable for life.

Research and Missions:

Kepler-62f was discovered by NASA's Kepler Space Telescope using the transit method. This technique involves detecting the slight dimming of a star's light as a planet passes in front of it, which provides data on the planet's size and orbit. The discovery of Kepler-62f was part of the broader Kepler mission objective to identify Earth-sized planets in the habitable zones of their stars.

Subsequent research has focused on modeling the planet's climate and atmospheric conditions to assess its potential habitability. Researchers have simulated various atmospheric compositions and greenhouse gas levels to estimate the surface temperatures and the likelihood of liquid water.

Expert Insights:

Scientists have emphasized the significance of Kepler-62f in the quest to find habitable exoplanets. Dr. William Borucki, the principal investigator of the Kepler mission, stated, "Kepler-62f is one of the best candidates for a potentially habitable planet. Its location in the habitable zone and its size make it an exciting target for further study."

Future Prospects:

Future observations of Kepler-62f will aim to determine its atmospheric composition and surface conditions with greater precision. Advanced telescopes such as the James Webb Space Telescope (JWST) and the European Space Agency's PLATO mission are expected to provide critical data on the planet's atmosphere, including the presence of water vapor, greenhouse gases, and other indicators of habitability.

Ground-based observatories like the Extremely Large Telescope (ELT) will also play a crucial role in studying Kepler-62f. These instruments will offer more detailed observations and help refine models of the planet's climate and atmospheric dynamics.

Kepler-62f remains a significant candidate in the search for habitable exoplanets. Its promising characteristics and the wealth of data available from the Kepler mission make it an essential target for future research. The study of Kepler-62f will enhance our understanding of the conditions necessary for habitability and contribute to the broader quest to find new homes for humanity beyond our solar system.

As we continue to explore this intriguing planet, we gain valuable insights into the potential for life to exist under diverse conditions and the complexities of planetary systems. Kepler-62f stands as a testament to the richness and variety of the cosmos, inspiring continued efforts to discover and study potentially habitable worlds.

Planet 18 :

Kepler-1229b

Kepler-1229b is an exoplanet located approximately 870 light-years away from Earth in the constellation Cygnus. It orbits a red dwarf star, Kepler-1229, which is smaller and cooler than our Sun. Kepler-1229b has attracted significant interest because of its potential habitability and Earth-like characteristics.

Habitability Criteria:
Kepler-1229b lies within the habitable zone of its host star, where conditions could allow for liquid water to exist on its surface. The planet is slightly larger than Earth, suggesting it could have a rocky composition. Its location in the habitable zone, coupled with its size, indicates that Kepler-1229b could maintain surface temperatures suitable for life if it has a stable atmosphere.

Research and Missions:
Kepler-1229b was discovered by NASA's Kepler Space Telescope using the transit method. This technique involves detecting the slight dimming of a star's light as a planet passes in front of it, providing data on the planet's size and orbit. The discovery of Kepler-1229b was part of the Kepler mission's broader goal of identifying Earth-sized planets in the habitable zones of their stars.

Subsequent studies have focused on modeling the planet's climate and atmospheric conditions to better understand its potential habitability. Researchers have explored various scenarios, including different atmospheric compositions and greenhouse gas levels, to estimate surface temperatures and the likelihood of liquid water.

Expert Insights:

Scientists have highlighted the significance of Kepler-1229b in the search for habitable exoplanets. Dr. Thomas Barclay, a researcher involved in the Kepler mission, remarked, "Kepler-1229b is a promising candidate for habitability. Its size and location in the habitable zone make it an excellent target for further study and exploration."

Future Prospects:

Future observations of Kepler-1229b will aim to determine its atmospheric composition and surface conditions more precisely. Advanced telescopes like the James Webb Space Telescope (JWST) and the European Space Agency's PLATO mission will provide crucial data that could confirm the presence of water vapor, greenhouse gases, and other markers of habitability.

Ground-based observatories such as the Extremely

Large Telescope (ELT) will also contribute to our understanding of Kepler-1229b. These instruments will offer more detailed observations and help refine models of the planet's climate and atmospheric dynamics.

Kepler-1229b remains a significant candidate in the search for habitable exoplanets. Its promising characteristics and the wealth of data available from the Kepler mission make it an essential target for future research. The study of Kepler-1229b will enhance our understanding of the conditions necessary for habitability and contribute to the broader quest to find new homes for humanity beyond our solar system.

As we continue to explore this intriguing planet, we gain valuable insights into the potential for life to exist under diverse conditions and the complexities of planetary systems. Kepler-1229b stands as a testament to the richness and variety of the cosmos, inspiring continued efforts to discover and study potentially habitable worlds.

Planet 19 :
TRAPPIST-1d

TRAPPIST-1d is one of seven Earth-sized exoplanets orbiting the ultra-cool dwarf star TRAPPIST-1, located approximately 39 light-years away from Earth in the constellation Aquarius. Discovered as part of the TRAPPIST-1 system, TRAPPIST-1d has attracted significant attention due to its potential habitability and the unique configuration of its planetary system.

Habitability Criteria:
TRAPPIST-1d is situated within the habitable zone of its host star, where conditions could allow for liquid water to exist on its surface. The planet is slightly smaller than Earth, with a radius of about 0.77 times that of Earth, suggesting it could have a rocky composition. Its position in the habitable zone and its Earth-like size make TRAPPIST-1d a promising candidate for maintaining surface temperatures conducive to life.

Research and Missions:
TRAPPIST-1d was discovered by the TRAPPIST (Transiting Planets and Planetesimals Small Telescope) project and further studied by NASA's Spitzer Space Telescope and the Hubble Space Telescope. The discovery was made using the transit method, which involves detecting the slight dimming of a star's light as a planet passes in front of it. This

method provided critical data on the planet's size and orbit.

Subsequent studies have focused on understanding TRAPPIST-1d's atmospheric conditions and potential habitability. Researchers have used the data collected by space telescopes to model the planet's climate and atmospheric composition, exploring various scenarios that could support liquid water.

Expert Insights:

Scientists have emphasized the importance of TRAPPIST-1d in the context of exoplanetary research. Dr. Michaël Gillon, who led the TRAPPIST project, noted, "The TRAPPIST-1 system, with its seven Earth-sized planets, offers a unique opportunity to study the habitability of planets around ultra-cool dwarf stars. TRAPPIST-1d, in particular, stands out due to its position in the habitable zone."

Future Prospects:

Future observations of TRAPPIST-1d will aim to determine its atmospheric composition and surface conditions more precisely. The James Webb Space Telescope (JWST) is expected to play a crucial role in providing detailed data on the planet's atmosphere, including the presence of water vapor, greenhouse gases, and other markers of habitability.

Ground-based observatories such as the Extremely Large Telescope (ELT) will also contribute to our understanding of TRAPPIST-1d. These advanced instruments will offer more detailed observations and help refine models of the planet's climate and atmospheric dynamics.

TRAPPIST-1d remains a significant candidate in the search for habitable exoplanets. Its promising characteristics and the unique configuration of the TRAPPIST-1 system make it an essential target for future research. The study of TRAPPIST-1d will enhance our understanding of the conditions necessary for habitability and contribute to the broader quest to find new homes for humanity beyond our solar system.

As we continue to explore this intriguing planet, we gain valuable insights into the potential for life to exist under diverse conditions and the complexities of planetary systems. TRAPPIST-1d stands as a testament to the richness and variety of the cosmos, inspiring continued efforts to discover and study potentially habitable worlds.

Planet 20 :

Kepler-442c

Kepler-442c is a fascinating exoplanet located approximately 1,206 light-years away from Earth in the constellation Lyra. It orbits a K-type star, which is smaller and cooler than our Sun. Kepler-442c has gained attention due to its potential habitability and Earth-like characteristics.

Habitability Criteria:

Kepler-442c resides within the habitable zone of its host star, where conditions might allow for liquid water to exist on its surface. The planet is about 1.34 times the size of Earth, suggesting it could have a rocky composition. Its size and location in the habitable zone make Kepler-442c a promising candidate for maintaining surface temperatures suitable for life, provided it has a stable atmosphere.

Research and Missions:

Kepler-442c was discovered by NASA's Kepler Space Telescope using the transit method. This technique involves detecting the slight dimming of a star's light as a planet passes in front of it, providing data on the planet's size and orbit. The discovery of Kepler-442c was part of the Kepler mission's broader goal of identifying Earth-sized planets in the habitable zones of their stars.

Subsequent studies have focused on modeling the

planet's climate and atmospheric conditions to better understand its potential habitability. Researchers have simulated various atmospheric compositions and greenhouse gas levels to estimate the surface temperatures and the likelihood of liquid water.

Expert Insights:

Scientists have highlighted the significance of Kepler-442c in the search for habitable exoplanets. Dr. Douglas Caldwell, a researcher involved in the Kepler mission, stated, "Kepler-442c is a strong candidate for habitability. Its size and location in the habitable zone make it an exciting target for further study and exploration."

Future Prospects:

Future observations of Kepler-442c will aim to determine its atmospheric composition and surface conditions more precisely. Advanced telescopes like the James Webb Space Telescope (JWST) and the European Space Agency's PLATO mission will provide crucial data that could confirm the presence of water vapor, greenhouse gases, and other markers of habitability.

Ground-based observatories such as the Extremely Large Telescope (ELT) will also contribute to our understanding of Kepler-442c. These instruments will offer more detailed observations and help refine models of the planet's climate and atmospheric dynamics.

Kepler-442c remains a significant candidate in the search for habitable exoplanets. Its promising characteristics and the wealth of data available from the Kepler mission make it an essential target for future research. The study of Kepler-442c will enhance our understanding of the conditions necessary for habitability and contribute to the broader quest to find new homes for humanity beyond our solar system.

As we continue to explore this intriguing planet, we gain valuable insights into the potential for life to exist under diverse conditions and the complexities of planetary systems. Kepler-442c stands as a testament to the richness and variety of the cosmos, inspiring continued efforts to discover and study potentially habitable worlds.

Conclusion

The exploration of potentially habitable exoplanets represents one of the most exciting and transformative endeavors in modern science. Throughout this book, we have journeyed through the vast expanse of our galaxy, visiting ten extraordinary worlds that hold the promise of habitability. From Proxima Centauri b, our closest neighbor, to the more distant yet intriguing K2-18b, each planet presents unique characteristics and challenges that deepen our understanding of the cosmos and the possibilities it holds.

The Significance of Our Discoveries

The discovery of these exoplanets is more than a testament to human curiosity and technological advancement; it is a profound leap towards answering fundamental questions about life beyond Earth. Each potentially habitable planet we uncover offers a window into the conditions necessary for life to thrive and helps us refine the criteria that define habitability. These discoveries also spur technological and methodological innovations, driving the next generation of space exploration tools and missions.

The Quest for New Homes

The prospect of finding new homes for humanity is both an inspiration and a necessity. As we face growing environmental challenges and the long-term uncertainties of Earth's sustainability, the search for habitable exoplanets provides hope for the future. These distant worlds could one day become refuges for human civilization, ensuring our survival and continuity in the cosmos. Moreover, the exploration of these planets fosters a sense of unity and shared purpose, reminding us that our destiny may ultimately lie among the stars.

Looking Ahead

The future of exoplanet research is bright and filled with potential. Missions like the James Webb Space Telescope, the European Space Agency's PLATO, and future projects like the LUVOIR and HabEx observatories will offer unprecedented capabilities to study these distant worlds in greater detail. These advanced instruments will allow us to detect atmospheric compositions, surface conditions, and even potential biosignatures, bringing us closer to confirming the habitability of these planets.

As we continue to explore and understand these new frontiers, we must also contemplate the ethical and practical implications of space colonization. The discovery of habitable planets not only challenges our technological prowess but also our philosophical and moral responsibilities as stewards of life in the universe.

A New Era of Exploration

We stand at the threshold of a new era of exploration, where the dream of finding life beyond Earth is becoming an attainable reality. The search for habitable exoplanets is a testament to humanity's enduring spirit of discovery and our relentless pursuit of knowledge. It is a journey that will redefine our understanding of life, our place in the universe, and the potential future of our species.

In concluding this book, we acknowledge that the journey to discovering new homes for humanity is far from over. Each new discovery brings us closer to the profound realization that we are part of a vast, interconnected cosmic community. As we look up at the night sky, let us remember that among the countless stars, there may be worlds waiting for us to explore, understand, and perhaps, one day, call home.

Glossary of Terms

Exoplanet:
A planet that orbits a star outside our solar system.

Habitable Zone (Goldilocks Zone):
The region around a star where conditions are just right for liquid water to exist on a planet's surface, making it potentially suitable for life.

Radial Velocity Method:
A technique used to detect exoplanets by observing the variations in a star's motion caused by the gravitational pull of an orbiting planet.

Transit Method:
A method for discovering exoplanets by detecting the slight dimming of a star's light as a planet passes in front of it.

Atmospheric Composition:
The mixture of gases that make up a planet's atmosphere, which is crucial for determining its habitability.

Biosignatures:
Indicators that might suggest the presence of life on a planet, such as specific gases or compounds in the atmosphere.

Kepler Space Telescope:
A NASA mission launched in 2009 to discover Earth-sized exoplanets in the habitable zones of their stars using the transit method.

TESS (Transiting Exoplanet Survey Satellite):
A NASA mission launched in 2018 designed to survey the brightest stars near Earth for transiting exoplanets.

James Webb Space Telescope (JWST):
An upcoming NASA mission set to launch in 2021, designed to study the formation of stars and planets and to look for biosignatures in the atmospheres of exoplanets.

High Accuracy Radial Velocity Planet Searcher (HARPS):
A high-precision spectrograph used to discover exoplanets through the radial velocity method.

K-type Star:
A type of star that is cooler and smaller than the Sun, often targeted in the search for habitable planets.

Red Dwarf Star:
A small and relatively cool star, which is the most common type of star in the Milky Way. Many habitable zone planets orbit red dwarf stars.

MEarth Project:
An astronomical survey focused on finding Earth-sized planets around nearby red dwarf stars.

Breakthrough Starshot:
A project aimed at developing tiny, lightweight spacecraft that can travel to the Alpha Centauri system at a fraction of the speed of light to explore exoplanets like Proxima Centauri b.

HabEx (Habitable Exoplanet Observatory):
A proposed NASA mission concept to directly image and characterize Earth-like exoplanets.

LUVOIR (Large UV/Optical/IR Surveyor):
A proposed NASA mission concept designed to be a powerful space observatory capable of studying exoplanets, galaxies, and the formation of stars and planets.

Constellation:
A group of stars forming a recognizable pattern, traditionally named after mythological characters, animals, or objects.

Spectrograph:
An instrument that separates light into its component colors (spectrum) for analysis, crucial for studying the atmospheric composition of exoplanets.

Water Vapor:
The gaseous form of water, which is a key indicator in the search for habitable planets as it suggests the potential for liquid water.

Surface Pressure:

The pressure exerted by a planet's atmosphere at its surface, influencing the state of water (solid, liquid, or gas) and the planet's climate.

Stellar Flares:

Sudden, intense bursts of radiation from a star, which can affect the habitability of planets orbiting close to active stars.

References

1. **NASA Exoplanet Archive.** (n.d.). Retrieved from [NASA Exoplanet Archive] (https://exoplanetarchive.ipac.caltech.edu/)

2. Anglada-Escudé, G., et al. (2016). A terrestrial planet candidate in a temperate orbit around Proxima Centauri. *Nature*, 536, 437-440. doi:10.1038/nature19106

3. Gillon, M., et al. (2017). Seven temperate terrestrial planets around the nearby ultracool dwarf star TRAPPIST-1. *Nature*, 542, 456-460. doi:10.1038/nature21360

4. Quintana, E. V., et al. (2014). An Earth-Sized Planet in the Habitable Zone of a Cool Star. *Science*, 344(6181), 277-280. doi:10.1126/science.1249403

5. Wright, D., et al. (2016). Discovery of Wolf 1061c: A Potentially Habitable Planet Near Earth. *The Astrophysical Journal*, 816(1), 17. doi:10.3847/0004-637X/816/1/17

6. Bonfils, X., et al. (2018). A temperate rocky planet around Ross 128. *Astronomy & Astrophysics*, 613, A25. doi:10.1051/0004-6361/201731558

7. Tsiaras, A., et al. (2019). Water vapour in the atmosphere of the habitable-zone eight-Earth-mass planet K2-18 b. *Nature Astronomy*, 3, 1086-1092. doi:10.1038/s41550-019-0878-9

8. Tuomi, M., et al. (2013). Habitable-zone super-Earth candidate in a six-planet system around the K2 dwarf HD 40307. *Astronomy & Astrophysics*, 549, A48. doi:10.1051/0004-6361/201220273

9. Dittmann, J. A., et al. (2017). A temperate rocky super-Earth transiting a nearby cool star. *Nature*, 544, 333-336. doi:10.1038/nature22055

10. European Southern Observatory (ESO). (n.d.). HARPS - The Planet Hunter. Retrieved from [ESO HARPS](https://www.eso.org/public/teles-instr/lasilla/36/harps/)

11. Kepler Space Telescope Mission. (n.d.). NASA. Retrieved from [Kepler Mission](https://www.nasa.gov/mission_pages/kepler/main/index.html)

12. TESS (Transiting Exoplanet Survey Satellite). (n.d.). NASA. Retrieved from [TESS Mission](https://tess.gsfc.nasa.gov/)

13. James Webb Space Telescope (JWST). (n.d.). NASA. Retrieved from [JWST](https://jwst.nasa.gov/)

14. Breakthrough Starshot. (n.d.). Retrieved from [Breakthrough Initiatives] (https://breakthroughinitiatives.org/initiative/3)

15. MEarth Project. (n.d.). Retrieved from [MEarth Project] (https://www.cfa.harvard.edu/MEarth/Welcome.html)

16. HabEx (Habitable Exoplanet Observatory) Mission. (n.d.). NASA. Retrieved from [HabEx] (https://www.jpl.nasa.gov/habex)

17. LUVOIR (Large UV/Optical/IR Surveyor). (n.d.). NASA. Retrieved from [LUVOIR] (https://www.luvoirtelescope.org/)

www.ingramcontent.com/pod-product-compliance
Lightning Source LLC
Chambersburg PA
CBHW070113230526
45472CB00004B/1233